The Macmillan Pre-Vocational Series

Series editor
Douglas Pride

Core Skills in Craft-based Activities

Contributors
David Draycott
Jeffrey Forbes
Derek Miles
Lawrence Miles
David Stevens
Terry Withington

Macmillan Education

Acknowledgements

The contributors and publishers wish to thank Judith Irving and Norman Smith for permission to reproduce Information Page A, pp. 62–7, taken from *Core Skills in Communication*; Burntwood Fasteners, Clive Guest, Trainwell YTS (West Bromwich) and staff at the Polytechnic, Wolverhampton, for their help with the photographs; and Christine Broome for typing the manuscript.

The contributors and publishers wish to acknowledge the following photograph sources: reproduced by kind permission of Angus Designs, p. 15; British Telecom Yellow Pages, pp. 7, 87; Central Office of Information, p. 61; Ford Motor Company Ltd, p. 68; HMSO, pp. 61, 71; Patricia Hamilton, p. 15; Andrew Nash, p. 16; National Westminster Bank PLC, p. 61; Charles Normandale, p. 15; Penguin Books Ltd from *The Alternative Printing Handbook* by Chris Treweek and John Zeitlyn, p. 34; Kate Rigby, p. 52; Sony (UK) Ltd, p. 46; Thomson Directories, p. 7; John Topham Picture Library, p. 46; Roger Viollet, p. 21; Derick Webster, pp. 10, 18, 19, 28, 42; T. Withington, pp. 8, 11, 31, 37.

The publishers have made every effort to trace the copyright holders, but where they have failed to do so they will be pleased to make the necessary arrangements at the first opportunity.

First published 1988

Published by
MACMILLAN EDUCATION LTD
Houndmills, Basingstoke, Hampshire RG21 2XS and London
Companies and representatives throughout the world

Printed in Hong Kong

British Library Cataloguing in Publication Data
Withington, Terry
Core skills in craft-based activities.
— (The Macmillan pre-vocational series).
1. Handicraft
I. Title
680   TT165
ISBN 0–333–42534–0

Contents

Introduction

This series is designed to help students meet the criteria laid down by the Joint Board for the new Certificate of Pre-Vocational Education. It aims to fulfil the three main requirements of CPVE courses:

(a) the integration of a core of basic skills with a wide-ranging choice of vocational studies;

(b) activity-based learning; and

(c) a flexibility which enables courses to be tailored to the needs of individual students.

The books in the series are arranged in two groups: one group concentrates on developing the main core competences using different vocational settings; the other concentrates on the skills required in the vocational categories (the CPVE introductory and exploratory modules), but also practises the core competences.

Each book consists of twenty assignments which develop skills in both general and specific vocational contexts. Ten of these concentrate on skills in general vocational contexts and ten on specific vocational situations (see diagram).

Main core competences				
Core Skills in Communication	Core Skills in Numeracy	Core Skills in Industrial, Social and Environmental Studies	Core Skills in Science and Technology	Core Skills in Information Technology
10 Core assignments (general vocational contexts) 10 Focus assignments (specific vocational situations) 10 Information pages				

Course

Vocational skills						
10 Introductory assignments (general vocational contexts) 10 Exploratory assignments (specific vocational situations) 10 Information pages						
Core Skills in Business and Administrative Services	Core Skills in Information Technology and Micro-electronic Systems	Core Skills in Service Engineering	Core Skills in Manufacture	Core Skills in Craft-based Activities	Core Skills in Distribution	Core Skills in Services to People

The assignments are free-standing and can be combined in different modular ways according to individual course needs. To assist selection and combination, the objectives of each assignment are given both at its head and in a grid at the beginning of each book. At the end of each book information pages give facts and advice to support the activities in the assignments.

Core Skills in Craft-based Activities contains twenty assignments, arranged in two groups: ten introductory assignments and ten exploratory assignments set in practical vocational contexts. The introductory assignments involve students in using the basic techniques and processes of craft-based activity, including planning and problem-solving, organising work, practical skills and the use of tools.

The exploratory assignments provide further opportunities for making and marketing products or monitoring workshop processes. Ten information pages provide guidance on key activities such as working in a team, health and safety, and behaviour in manufacturing workshops.

The material in this volume is also suitable for a wide range of other programmes, such as YTS off-the-job training, and vocational preparation and pre-vocational initiatives, including the BTEC/City and Guilds Foundation Programmes.

CPVE grid

Assignments	Personal and Career Development	Industrial, Social and Environmental Studies	Communication	Social Skills	Numeracy	Science and Technology	Information Technology	Creative Development	Practical Skills	Problem-solving
1 What can we make?	■	■	■	■				■	■	■
2 Hazard warning	■	■	■	■		■			■	■
3 Compelling selling	■	■	■	■			■	■	■	
4 Tools of the trade	■	■	■			■			■	■
5 Custom-built	■		■	■				■	■	■
6 Making paper	■	■				■		■	■	■
7 The box for the job	■	■	■		■			■	■	■
8 Job shadowing: preparation	■	■	■	■			■			
9 Bill's Bikes	■	■	■		■				■	■
10 Silk-screen workshop	■		■	■				■	■	■
11 The identity tag		■	■			■	■		■	■
12 Job shadowing	■	■	■	■						
13 High-tech help	■	■	■				■	■	■	■
14 Housework	■	■		■						■
15 Second edition	■	■	■				■	■	■	
16 Gummed up		■	■			■		■	■	■
17 Work-experience project: costings	■		■		■				■	■
18 Convenience foods	■	■		■	■				■	■
19 On the safe side	■	■	■	■	■					■
20 Setting up in business	■	■	■	■	■				■	

Introductory assignments

1

What can we make?

──── AIM ────

To develop your understanding of
- the skills required in craft-based activities
- the materials and processes needed to set them up
- planning and organisation

Introduction

Most craft production uses skills and understanding which many people may already have, but which need to be practised and improved on. Today many people in the United Kingdom make a living by setting up small craft workshops which use skills that they have developed through a leisure interest or hobby.

This assignment will help you to find out what skills are used to make simple products and also to help you find out what skills you and others have. It will also give you practice in using and improving these skills.

Task 1

Choose a product which you think you and perhaps others in your group could make reasonably well with the skills you have *now*. It may help to brainstorm a list of ideas. List on a sheet of paper all the simple products you use around school or college and at home that you might be able to make. From your list, choose *one* that you would like to make.

Task 2

You will now need to find out how to make your product. One interesting way is to find someone who is already making it — look through *Yellow Pages* or contact local craft shops or the local chamber of trade. Arrange a visit to see the product being made.

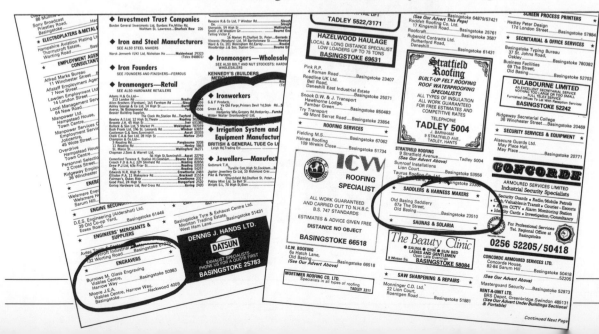

Before going on the visit, make a list of what you need to find out. The following list may be helpful:
(a) Materials needed to make this product.
(b) Tools needed.
(c) Is it made in parts? (List them or make a rough sketch.)
(d) Is it made in stages? (List them or make sketches.)
(e) How is it finished off?

Task 3
Now list all the skills needed to make the chosen product. Having seen someone making it on your visit you should already have a good idea of the skills needed, but it will be useful to list them as clearly as possible. Look at the example in Figure 1.

Figure 1 List of skills needed

> **Stripping paint from a door**
> 1 Working safely when using corrosive materials
> 2 Working safely when using a blowlamp
> 3 Organising work
> 4 Removing and refitting hinges and locks
> 5 Using a paint scraper
> 6 Removing paint from mouldings with a small knife
> 7 Mixing and using sugar soap
> 8 Sanding doors to a smooth finish

Task 4

Now that you know which skills you need to make your product, form a production team. Using the list from Task 3, design and distribute a questionnaire to find out which students in your group have the skills you need.

Having identified those students who have these skills, discuss and negotiate with them the parts they would play in the production team.

Task 5

Finally, write a *feasibility study* for your chosen product — that is, a report showing what would be involved in going into full production, and mentioning any snags you foresee.

Note: Assignment 5 develops the work done here.

_____IMPORTANT_____

Read these information pages:
A Social-survey techniques
G Working in a team

2 Hazard warning

___AIM___

To develop your
- knowledge of health and safety regulations
- understanding of how they are applied
- ability to identify and record processes and methods
- ability to plan an investigation and record conclusions

Introduction

An important aspect of work in a craft workshop is knowing the health and safety regulations, and making sure they are acted upon. These regulations apply to the physical conditions and to the machinery, tools and material used in the production process. Acids are used in certain metal treatments; dyes are applied to leather and fabrics; and blacksmiths, for instance, require very high temperatures to carry out their work. Fumes are given off when chemicals are used or bonding agents applied, and dust is created in heating processes. Proper precautions must be taken in each case.

This assignment is designed to help you to understand more about these regulations. There is also a task which relates to accident procedures. The assignment can be an individual piece of work relating to a particular craft-based activity which is of interest to you, or a group activity in which together you investigate a number of craft-based areas. The work can be carried out on your own site, in a commercial craft-based workshop, or as part of your work-experience programme.

Task 1

Arrange a visit to a craft-based workshop, on-site or off-site. First decide who is the appropriate person to contact, the person who can give permission for you to visit. You will need to state:

(a) who you are and where you are from;

(b) the purpose of your study;

(c) the most convenient time for your visit or visits (you may need more than one).

Task 2

To prepare for your visit you will need to devise a work schedule. Read through all the tasks in this assignment before you start planning. Tasks 3–7 give you specific areas of investigation, and will help you to organise your investigation and your report.

Task 3

Record:

(a) the craft process or processes that are taking place in the workshop (there could well be more than one);

(b) the machinery and tools that are being used;

(c) any treatment processes that require the use of chemicals, acids, dyes, etc.

Task 4

Find out:

(a) the safety regulations that apply to the processes, machinery, tools and treatments;

(b) the source of these regulations (e.g. the Health and Safety at Work Act, Local Education Authority codes of practice, or local agreements made between employers and trade unions);

(c) the person(s) responsible for the observance of these regulations.

Task 5

Find out:

(a) How are the safety regulations communicated to students and employees (e.g. by written instructions, signs or notices, or by word of mouth)?

(b) Is there a health and safety induction programme? If so, when does it take place? Who is responsible for the programme?

(c) Are there fire drills? How often do they take place?

Task 6

In the event of an accident:

(a) What are the procedures?

(b) Is there a trained first-aider? Do people in the workshop know who he or she is?

(c) Is there a first-aid box? Is it properly equipped?
(d) How are accidents formally reported?
(e) What are the most common accidents? Why?
(f) What steps have been taken to reduce these sorts of accident?

Task 7
Compile a report of your investigations. Include photographs or diagrams if you wish to illustrate and draw attention to issues which you think are important. What conclusions have you come to as a result of your study? Have you identified areas which could be improved? Have you any ideas for improvements? Remember, for your conclusions to be valid you must base them on the evidence you have collected during your study.

_____IMPORTANT_____

Read this information page:
C Health and safety

3   Compelling selling

To develop your
- ability to identify the basic elements of marketing
- awareness of the role of marketing in small companies

MONTHLY SALES FIGURES

SUCCESS in BUSINESS

UP YOUR BUSINESS

BUSINESS & success

Introduction

No matter how good a product or service is, a business cannot succeed unless the views and needs of the customers are taken into account.

Some craftspeople are surprised when customers do not buy their products. They know that the quality and skill involved in the manufacture is of a high level; they expect those who are spending money to prefer their own product to one that is similar and cheaper but produced by a machine.

The whole process of presenting and promoting the product or service to the customer is called *marketing*. It is an important ingredient in a firm's success. This assignment gives you the opportunity to learn more about it.

Task 1

Selling goods and services is more complicated than many people think.

Consider a craftswoman, working alone, who makes cane baskets. At the present time she makes log baskets and waste-paper baskets. Sales are made from the workshop where the work is carried out.

Copy the table in Figure 1. For each section, fill in the three examples. Then describe the advantages and disadvantages of each.

Figure 1 Ways of developing a business

	Advantages	*Disadvantages*
Other ways of selling *e.g. mail order* 1 2 3		.
Possible new products *e.g. shopping baskets* 1 2 3		
How to increase sales *e.g. reduce prices* 1 2 3		
How to increase profits *e.g. use cheaper materials* 1 2 3		

Task 2

Marketing is not just about customers paying money for products. In today's fast-moving and competitive world there is a great deal more to it. To complete this task you will need to do some research both in a library and by talking to the sales staff of firms in your area.

Figure 2 is a list of marketing terms. Find out what they mean. Make a list with two columns: the marketing terms down the left-hand side and clear explanations on the right.

Figure 2 Marketing terms

Market research	Point of sale	Mail-order sales
Product trials	Product launch	Trade fairs
Customer profile	Competition	Closing a sale
Packaging	Servicing policy	Displays
Potential market	Promotion	Retail sales
Pricing policy	Direct mailing	Product range
Sales literature	Image building	Customer feedback
Product features	Party plan	

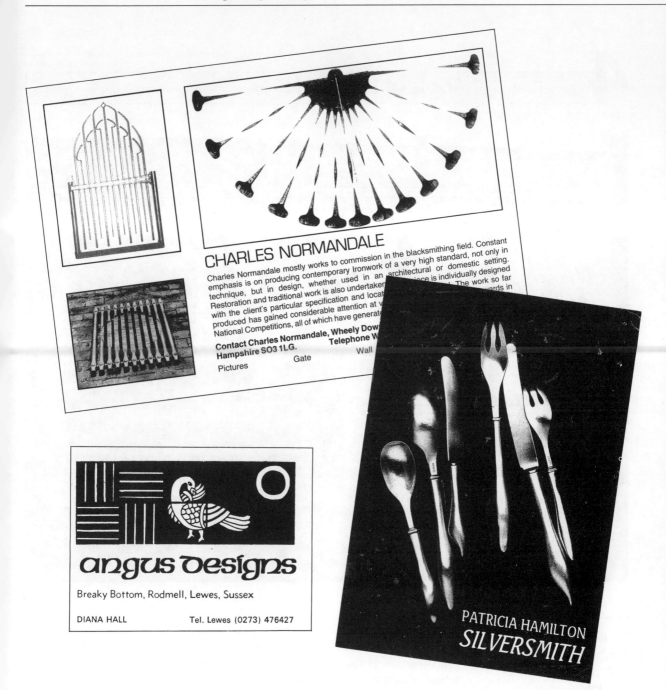

CHARLES NORMANDALE

Charles Normandale mostly works to commission in the blacksmithing field. Constant emphasis is on producing contemporary Ironwork of a very high standard, not only in technique, but in design, whether used in an architectural or domestic setting. Restoration and traditional work is also undertaken. Each piece is individually designed with the client's particular specification and location. The work so far produced has gained considerable attention at various awards in National Competitions, all of which have generated...

Contact Charles Normandale, Wheely Dow... **Telephone W...**
Hampshire SO3 1LG.
Pictures   Gate   Wall

angus designs

Breaky Bottom, Rodmell, Lewes, Sussex

DIANA HALL   Tel. Lewes (0273) 476427

PATRICIA HAMILTON
SILVERSMITH

—— IMPORTANT ——

Read this information page:
J Sources of information

Task 3
Using the information you have obtained about these marketing activities, write a marketing plan explaining the ways in which the basket maker might develop her sales over the next year. Where you suggest an activity, such as advertising, give an estimate of the cost involved and the likely benefits to the business.

4 Tools of the trade

AIM

To develop your
- ability to make freehand sketches of tools
- knowledge of the use of specialist tools

Introduction

Throughout history craft skills have been kept secret by the crafts-people themselves. When visitors entered the workshop workers often put down their tools. There are some tools now exhibited in museums whose exact use we do not know – only those workers skilled in their use could show others how to use them.

This assignment should ensure that details of at least a few modern tools will be preserved!

Task 1

Select three tools in current use. Present a report on each tool, using the example in Figure 1. Include a sketch of the tool.

**Figure 1 A sample
report on a tool**

Tool Revolving head punch.

Size (overall) About 300 mm.

Punch diameter 2−5 mm.

Material Steel.

Use To punch holes in leather or such materials.

Method of use A punch is selected by revolving the head to give the correct size of hole. The material is positioned between the jaws, which are then clamped together.

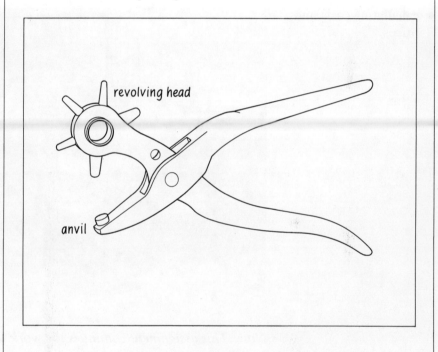

Task 2
On the sketches, identify and label the main parts.

Task 3
For each tool, identify the items most likely to wear in normal use. List these. State whether the item is replaceable when worn; if it is, give the approximate cost.

_____IMPORTANT_____
Read this information
page:
F Care of tools

Task 4
Select one of the tools included in your report and describe how this tool may have been improved over the years to make it more useful, safer and more efficient.

5 Custom-built

——— AIM———

To develop your skills in
- designing and making simple products
- working in a group
- calculating costs and prices

Note: This assignment continues the work done in Assignment 1.

Introduction

Custom-built goods are products that are designed and made to meet the special needs of particular customers. *Customised goods* are mass-produced items that have been adapted to meet the needs of particular customers. The racing bike of a professional cyclist is likely to be custom-built; if your bike has had drop handlebars or extra lights fitted since it was bought it has been customised.

In either case the attractiveness of the product has a lot to do with the fact that it meets *individual* needs. Many products, especially those that are 'hand-made', are marketed on the basis that they meet very specific individual requirements. To take a grand example, individual architect-designed houses are relatively expensive. This is not because they are better built, and they may not even be more attrac-

Figure 1 Possible products in three craft areas

Knitting cardigans jumpers dresses shawls **Woodworking** tables tool boxes garden furniture shelf units **Pottery** vases, bowls cups, mugs plates lampstands

tive: the buyer is paying for the individual attention, for a product that is exactly tailored to his or her needs. On a smaller scale, tailor-made suits can command a higher price partly because they are made to fit individuals.

This assignment gives you an opportunity to look closely at the idea of producing something that the customer wants — what this means for a small craft-based industry, and what the difficulties are. You will need to operate in groups of 3–5 for the three tasks. Wherever possible, make up groups on the basis of your own shared interests.

Task 1
Look at the list of craft-based activities which you identified in Assignment 1. Select *one* craft activity for your group — perhaps the one you chose for Tasks 2–5 of Assignment 1. Remember that you need to select an activity for which your group has the necessary skills.

Once you have chosen the craft activity, select *one particular product* which you could make in your school/college or at home. The list in Figure 1 may give you some ideas of products in three areas.

Task 2

You wish to tailor this product to an individual need, so you have now to find a single interested customer. All members of the group need to collaborate on this; pool your resources, both time and possible customers!

Having identified a customer, discuss his or her exact requirements. Make a checklist of what is wanted. This should *at least* include:

(a) the main design features;
(b) the material(s) to be used;
(c) the date the item is required;
(d) the approximate cost.

Task 3

Design and make the article according to the checklist. This may well involve a number of stages, such as:

(a) discussing as a group what needs to be done, and who does what;
(b) negotiating with your tutor on access to equipment;
(c) agreeing a programme of work;
(d) production.

One of your group should be appointed 'monitor' to check that the work is being done to the specification.

Task 4

How can you check that the customer is satisfied? Go through your original checklist with the customer and check off the requirements that have been met.

Are there any items on the list that *cannot* be ticked? Are there good reasons for this? Write brief notes (250 words maximum) on this production project. Highlight any problems in design, production or marketing that you came up against and how you tried to overcome them.

———IMPORTANT———

Read these information pages:
B Better by design
I   Costs

6 | Making paper

AIM

To develop your
- ability to use simple equipment and materials
- ability to plan and perform a range of simple craft-based activities

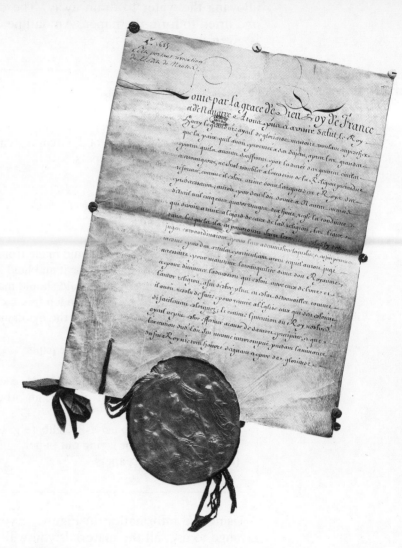

Introduction

Paper can be made by large-scale industrial processes using equipment worth millions of pounds and employing thousands of workers. It can also be made using simple equipment found in most homes and small craft workshops.

In this assignment you will become familiar with the process of making paper and learn the relevant skills. In doing so you will set up your own paper-making facility, and practise planning and carrying out a simple craft-based activity.

Task 1

Paper-making is a simple process. It consists of mixing vegetable fibres − usually using waste paper as the basic raw material − with water to make a pulp, and then spreading this pulp evenly and allowing the water to drain away. The drained pulp is then pressed and dried to form the paper. An outline of the paper-making process is given in Figure 1.

Figure 1 Stages in the paper-making process

Stage 1: Making the pulp
Equipment and materials needed:
(a) A large bucket.
(b) A smooth piece of wood, 50 cm × 4 cm × 4 cm.
(c) Plenty of old newspapers.
(d) A supply of water.

Stage 2: Moulding the paper
Equipment needed:
(a) A large sink.
(b) A rectangular paper mould made from 4 cm × 4 cm wood, with silk mesh stretched across it and held by staples. The internal measurements of the mould should match an A4 sheet of paper.
(c) A wooden *deckle* − a wooden frame exactly the same size as the paper mould, but without the nylon mesh.

Stage 3: Pressing and drying the paper
Equipment needed:
(a) 2 rectangular pieces of hardboard, each slightly larger than A4.
(b) 25 cloths made from smooth, absorbent material, again slightly larger than A4.
(c) Heavy objects to act as weights (e.g. bricks).
(d) A flat surface for laying out paper.
(e) More old newspapers.

Using the information in Figure 1, your first task is to obtain, or arrange to use, all the materials you will need. Remember to arrange to use a suitable workshop.

Task 2

Now make the moulding screen and deckle. Measure a piece of A4 paper and calculate the dimensions of the moulding screen. (Remember that its *internal* measurements should be the size of sheet of A4 paper.) Make two wooden frames; one is the deckle (Figure 2). Stretch the nylon mesh tightly across the second frame, using a staple gun to secure it. Make sure it is stretched evenly and tightly.

Figure 2 Paper mould and deckle

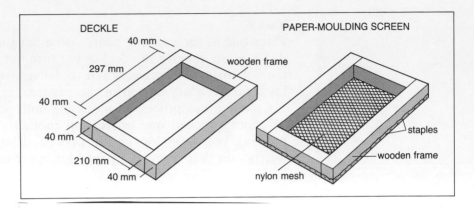

DECKLE

PAPER-MOULDING SCREEN

40 mm

297 mm

wooden frame

40 mm

40 mm

210 mm

40 mm

staples

wooden frame

nylon mesh

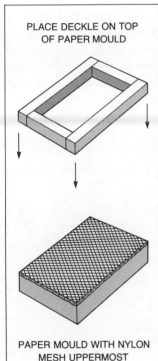

PLACE DECKLE ON TOP OF PAPER MOULD

PAPER MOULD WITH NYLON MESH UPPERMOST

Figure 3 Moulding paper

Task 3

Make the pressing boards. Make sure that they are slightly *larger* than the internal dimensions of the paper mould, as they must completely cover the paper you make.

Task 4

You are now ready to make the pulp. Begin by tearing up 25 to 30 full sheets of old newspaper. (Large-format newspapers such as *The Guardian* should be used for this.) Place the torn paper in the large bucket and add water sufficient to soak all the paper thoroughly. Leave this mixture for 1 hour.

Beat and stir the mixture with the wooden stick until the paper is completely broken down into its basic fibres. You may need to add more water to do this, but only add enough to make the pulp slushy. To test the pulp, take some out in a jar and add twice as much water. If you cannot see any recognisable particles the pulp is ready for the next stage. If you can still see particles you will have to carry on beating the pulp.

Task 5

This is the moulding stage of the paper-making process.

First pour the pulp into the sink and add about 11 litres of water. It is useful to keep some of the undiluted pulp back in case you need to thicken the pulp in the sink later. Stir the pulp in the sink frequently throughout the moulding stage so that it does not settle.

To mould the paper, place the deckle against the mesh side of the moulding screen (Figure 3). Holding them both together, lower them into the pulp mixture, mesh side uppermost. Carefully hold the mould and deckle just under the surface of the mixture, pivoting the mould slightly. This will lay a coat of pulp over the mould – it should be about 3 mm thick.

When you have done this, gently take the mould and deckle out, allowing the pulp to drain. At this stage a slight rocking movement will help the pulp to settle evenly on the mould.

Task 6

Place one of the pressing boards on a flat surface and lay a cloth on it. Carefully remove the deckle and turn out the sheet of paper onto the flat surface, taking care not to damage the sheet of paper. Then lay a further cloth on top of the paper.

Continue this process of moulding until you have about 20 sheets of paper, stacked one on top of another. Then place the second pressing board on top of the pile of sheets and gently place the weights on that. Press the paper for approximately 10 minutes.

Task 7

When the paper sheets have been pressed for 10 minutes, carefully remove them one by one and lay them to dry on a flat surface covered with old newspaper. Leave the cloth in place at this stage.

After about an hour turn each sheet over onto fresh newspaper to allow the other side to dry. Remove the cloth at this stage. When both sides are thoroughly dry your hand-made paper is ready for use.

_____IMPORTANT_____

Read this information page:
G Working in a team

7 The box for the job

To develop your
• understanding of a design problem
• ability to express facts and information
• ability to interpret visual information

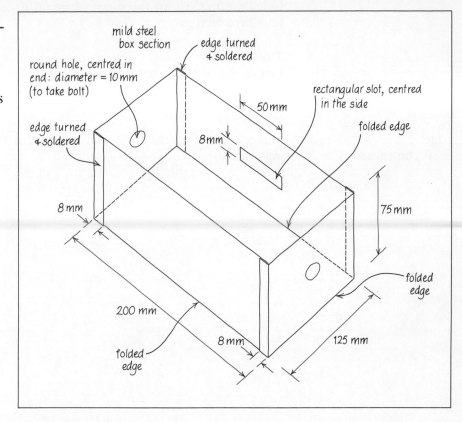

Figure 1 Sketch of a box section

Introduction

The sketch in Figure 1 shows a box section required for a boiler in a heating system.

This kind of design often starts on the back of an old piece of paper. The engineer works out what is needed to overcome a design problem and makes a simple sketch of the item that provides the answer. The sketch in Figure 1 is an example. It has been given to your group; you are asked to provide a detailed drawing and to specify the method of production, so that someone else can make it.

The box section shown above is one part of a heating system. Four such units are required for each complete system. In use, the rectangular slot is fitted over a tab; the tab holds very little weight. The two holes, one at each end, hold a bolt which travels through the box section. The box itself is made from small-gauge mild steel and will be spot-welded at the base to a larger section of the boiler.

Task 1
Draw the box section as it would appear when dismantled, as a flat piece of metal. Include on the drawing all the dimensions necessary.

Task 2
Calculate the size of the piece of metal necessary to manufacture all *four* box sections.

Task 3
List in detail the processes necessary in the manufacture of the box section. Ensure that the steps are presented in a logical order.

_____IMPORTANT_____

Read this information
page:
E Technical terms

Task 4
Make a list of all the tools and equipment that will be needed in its manufacture.

8 Job shadowing: preparation

_____AIM_____

To develop your skills in
- identifying the activities involved in particular skills
- working effectively in groups
- communicating to a particular audience

Introduction

You will be able to remember occasions when you have watched a skilled craftsperson at work, perhaps when you were at a craft fair or making an industrial visit. The combination of the skills, the confidence, the speed and the unfailing accuracy will have impressed you. Perhaps when you have watched a skilled craftsperson in an area of interest to your vocational studies you have felt your own confidence drain away: 'I'll never be able to do that!' Yet the people you have watched were once trainees, students or apprentices, and they had to start at the beginning and develop the skills they now have.

This assignment provides you with an opportunity to find out about the skills and personal qualities that a craftsperson requires, and to arrange a meeting with someone.

Task 1

Discuss with the others in your group what various craft jobs may involve. Make a list, as shown in Figure 1. Write down the activities that a person doing the job might reasonably be asked to do. Alongside, list the skills that the craftsperson would need in order to carry out those activities successfully. Figure 2 shows an example.

Figure 1 Job skills needed by a craftsperson

Job:	
Activities	*Skills*

Figure 2 A sample list of skills

Job: JOINER (BENCH)	
Activities	*Skills*
Cuts, shapes and fits wooden parts in workshops to form structures and fittings ready for installation on site.	1 Measuring accurately. 2 Selecting correct timber. 3 Organising work to meet deadlines. 4 Cutting, shaping and assembling accurately. 5 Following specifications.

Task 2

When a *financial audit* is carried out in business, the records of work done, costs and payments, and the holdings of stock and equipment are all checked thoroughly. Similarly, a *skills audit* looks at the whole range of skills used by a craftsperson. Employers and many agencies need to know about jobs in this detail in order to match the right person to the job.

Job Centres and the Careers Service make use of a large reference book called the *Classification of Occupations and Directory of Occupational Titles* − or *CODOT*, for short. It is published by the Department of Employment, and attempts to classify all occupations according to the skills and knowledge involved.

Arrange to see this. You could try a Job Centre, the Careers Service, the local office of the Training Commission or a main public library's reference department.

Task 3

In small groups, make a shortlist of a number of craft-based jobs that interest you, using the *CODOT* index for reference and information where necessary. Allocate one job to each person.

Your objective now is to make contact with someone who *does* that job, so that you can find out more about what he or she actually does. Using the Careers Service, Job Centres, a local business directory or group knowledge, identify a company that may be employing someone to carry out that particular job.

———IMPORTANT———

Read this information
page:
G Working in a team

Task 4

Write a letter to the employer identified. Outline the course you are following and the purpose of the enquiry. Ask permission to 'shadow' the person doing this job for one day. 'Shadowing' as used here means following a person throughout the working day, recording everything he or she does in order to form a detailed picture of the job.

Note: The job shadowing itself is carried out in Assignment 12.

9 Bill's Bikes

To develop your
- understanding of the marketing process
- skills in collecting and presenting information
- skills in calculating and comparing costs
- ability to work effectively in a group

Introduction

Large companies or public organisations often have whole departments working exclusively on marketing, advertising or promoting their products. This indicates the importance employers now attach to the business of presenting themselves and their goods effectively to their potential clients. Such is the scale of marketing activity that people often now refer to the 'advertising industry'. There are companies specialising in market research, in public relations, in graphic design, in television promotion, in public relations and a whole range of other aspects of advertising.

Small craft-based businesses with two or three employees rarely have the resources to pay for their own publicity and promotions department or to go outside for advertising expertise. But the job of marketing is just as important to small businesses, and needs to be considered seriously.

This assignment gives you the experience of working through the kinds of problems which a small company encounters in trying to promote itself. It should also show you how the size of a company affects the way in which it operates.

Task 1

Read Figure 1.

Bill and Jane have to start by looking at a variety of ways in which they could advertise. In small groups, brainstorm a number of ways that you know of which small companies use to promote themselves. Elect one of the group to write the methods down. Think of as many ways as possible *in 10 minutes*.

Task 2

Now refine your list of possibilities by looking more closely at what small companies in your area actually do.

Within your group, allocate to each individual a particular kind of small — preferably craft-based — business which operates in your area. You might consider:
- small bakers
- tailoring or garment repair
- fly-fishing accessories
- bicycle repair
- furniture renovation and repair
- ornamental wrought-iron work

Figure 1 A case study

> **Bill's Bikes**
>
> *Bill's Bikes* is run by Bill Davenport and his daughter, Jane.
>
> When he was made redundant, Bill decided to develop an existing interest in bicycles into a business. Over eighteen months he built up a small enterprise, repairing and refitting bicycles. Jane, a keen amateur cyclist, joined her father in the business at the end of her time on a Youth Training Scheme, during which she worked in an estate agent's office.
>
> Most of the business has been brought in by word of mouth; both Bob and Jane have contacts in amateur cycling and road-racing. People with expensive bikes are pleased to have repairs done by people they know and trust.
>
> However, it is obvious to both Bill and Jane that more could be done to make *Bill's Bikes* better known — business is literally cycling by the door every day. They have checked their accounts, and estimate that they could spare around £500 a year to market themselves.

Each person then needs to identify 2—6 businesses working locally in the chosen area. You could find these by looking through the classified section of your local newspaper, the 'small ads' display in your local supermarket, *Yellow Pages* or a local industrial directory. Just by looking for information in these places, you will actually have identified some of the ways in which businesses advertise themselves.

Task 3

Devise a simple questionnaire to identify the ways in which these businesses advertise and how much it costs them. Use the list from Task 1 as the basis of the questionnaire.

Now use the questionnaire. *Either* use it as the basis for a face-to-face interview with employers *or* send it to employers, with a covering letter explaining what you are doing and why. In either case, ask also for samples of advertising material. Set a deadline for return of these questionnaires, so that the whole group can arrange a time to meet again and compare results.

Task 4

In your groups, discuss the 'feedback' that you have had from your questionnaires. It will be interesting to discuss:

(a) whether different types of small business use the same kinds of advertising;
(b) whether the questionnaires revealed kinds of advertising which you weren't aware of;
(c) whether some questionnaires were better than others at getting information.

Task 5

Bearing in mind what you know about *Bill's Bikes* – including the marketing budget – produce a short report for Bill and Jane outlining some of the ways of advertising themselves which they ought to consider. Say how much some of these ways would cost. Include a particular worked example of a low-cost advert (such as a small advertisement in a supermarket or a publicity handout) for *Bill's Bikes*.

——————IMPORTANT——————

Read these information pages:
A Social-survey techniques
J Sources of information

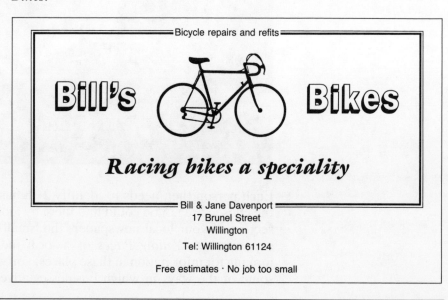

10 Silk-screen workshop

AIM

To develop your
- skills in organising materials and components for simple craft-based activities
- skills in designing and constructing simple equipment
- ability to evaluate the cost of materials and equipment

Introduction

Setting up many simple craft-based activities involves similar skills and procedures. For example, identifying the necessary tools and materials and how you can design and then contruct the simple equipment needed are a part of all craft-based processes. In completing this assignment you will learn how to organise your work by breaking it down into simple stages, you will collect together the materials nccdcd to set up a silk-screen workshop, and you will construct silk-screening equipment and use it to make something of saleable quality.

Task 1

Read Figure 1. Then examine the drawing in Figure 2. It shows you the basic equipment needed to set up your own silk-screen workshop.

Figure 1 Silk-screen printing

Silk-screen printing is a way of printing a design onto paper or material. Coloured dyes are squeezed through a stencil, the stencil being held in place by a screen. The process is called 'silk-screen printing' because the screens were originally made of silk.

Figure 2 Silk-screening equipment

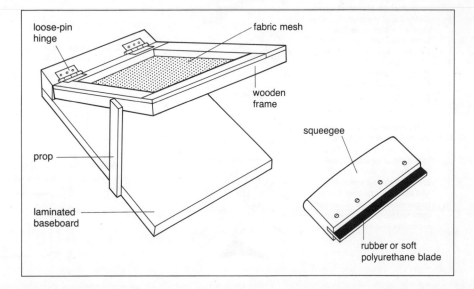

Figure 3 Fixing the mesh to the screen

To tape up the screen. You will need:
- a roll of gummed tape (at least 5 cm/2 in. wide).
- a bowl of water.

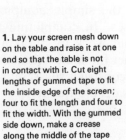

1. Lay your screen mesh down on the table and raise it at one end so that the table is not in contact with it. Cut eight lengths of gummed tape to fit the inside edge of the screen; four to fit the length and four to fit the width. With the gummed side down, make a crease along the middle of the tape by folding it in half all along its length. Do this to two pieces of tape of each length.

2. Dip the tape right into the water, lift it out and get rid of the excess water by holding the tape between two fingers and running them down its length.

3. Stick the tape into the right angle between the mesh and the frame, stretching it slightly as it is being stuck so that it does not pucker. Stick a second length of tape all round on the flat of the mesh, overlapping on to the first piece. Stick two more strips of tape across the top and bottom of the screen to make the well into which the ink is poured and where it sits between printings.

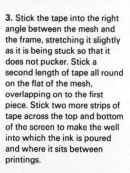

4. Reinforce the corners with small pieces of tape, folded and cut as shown. If you are going to use water-based inks, paint the tape with shellac or varnish to stop the water dissolving it. (An easier and quicker way is to use masking tape instead of gummed tape, but this is more expensive and not as reliable.)

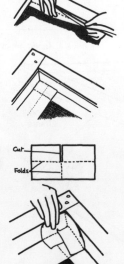

Stretching by hand

1. Cut a piece of mesh about 15 cm (6 in.) larger than the outside size of the frame. Wet it with water (it stretches more easily wet).

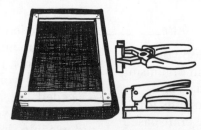

2. Lay the fabric in position across the frame, making sure that the threads are running at right angles to it.

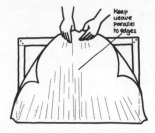

3. With a staple gun, staple the mesh to one of the short sides, starting at the middle and working out to the edges.

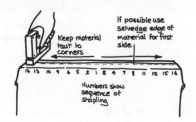

4. Staple the opposite edge in the same way, making sure that you pull outwards and sideways very tightly.

5. Repeat steps **3** and **4** on the two remaining sides to obtain a screen which is drum-tight with no puckers or unevenness in the mesh (staple down the overlap at the corners).

You will also need

(a) sheets of newspaper;

(b) a sharp modelling knife, to make the paper stencils;

(c) water-based printing inks of various colours;

(d) paper to print your designs on;

(e) a staple gun, gummed tape and clear varnish, to fix the silk to the screen frame.

Your first task is to design and draw the equipment you are going to make. Show all the dimensions. For your first attempt it is advisable not to make your silk screen too large.

Task 2

Having decided on all the dimensions for your silk screen, write a list of all the tools and materials you need to make it. Then find out which tools and materials are already available in your craft workshop, and which it will be necessary to buy.

Next you will need to find suppliers of whatever materials and equipment you need, and get estimates of costs. Present a written report to your tutor or group on the costs involved in setting up the silk-screen workshop.

Task 3

Working closely with your tutor, arrange for the purchase of the materials you need. Wherever possible you yourself should keep the financial records, make the telephone calls and write all the letters necessary in making the purchases.

Figure 4 A simple stencil

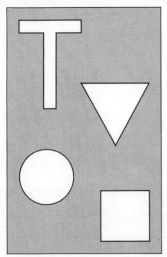

Task 4

Having obtained all the tools and materials you need to make the silk screen, consult with your tutor about use of the craft workshop or wood workshop to make the screen.

You will need to plan your work carefully before making the screen and base board. Figure 3 shows you how to stretch and fix the nylon mesh on the screen. It is important to get this right: the tightness of the screen will affect the quality of the print.

Task 5

You are now ready to make your stencil (Figure 4). It is better to choose a simple design − perhaps just basic shapes − for your first attempt.

Using a blank sheet of paper the size of your screen, draw your design and then using the modelling knife cut it out. Remember that *what you cut out is what you print*. Make sure your design is in the centre of the sheet.

Task 6

Now print your first design. First choose the colour you want to use and mix the ink. Follow the instructions supplied, but make sure the ink is not too thin.

Place a sheet of paper on the bed of the silk screen and position your stencil on top of the sheet of paper. Lower the screen so that it is flat on the bed. Now pour a small amount of ink onto the edge of the screen away from you. To make a print, pull the squeegee across the screen at an angle of 45 degrees, drawing the ink with it and forcing the ink through the nylon mesh onto the printing surface. Make sure that the ink is drawn across the screen evenly, with no streaks − if there *are* streaks, pull the squeegee across the screen again.

Now lift the screen up from the bed and set it on the prop. You will see that the stencil is firmly attached to the screen. Your print may also be attached to the screen! If so, gently pull it away, *taking care not to remove the stencil from the screen*. Place your print in a safe place to dry.

You can now repeat this process to make more copies of your design. Paper stencils will often make up to 50 prints from the same stencil.

———IMPORTANT———

Read these information pages:
B Better by design
J Sources of
   information

Exploratory assignments

11

The identity tag

—————AIM—————

To develop your:
- understanding of market research
- ability to design a questionnaire
- ability to carry out small-scale market research
- ability to present recommendations
- skills in estimating and costing tasks

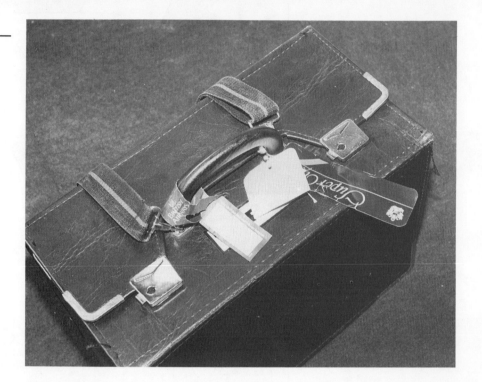

Introduction
In this assignment you are going to carry out some *market research* — into which kind of identity tag students would like to put onto their bags or cases. You will report your findings as if you had been requested to carry out the task for a manufacturer. This kind of market research is standard practice in business: the manufacturer wants to be sure that the final product is something that people really will buy.

There are many ways of carrying out market research; for this assignment an interview technique with a questionnaire will probably be suitable. (This assignment would more easily be carried out by a group, therefore.)

Task 1
(a) Think first about the *design* of the identity tag. Will it carry any wording? Or an illustration?
(b) Now think about the *shape* of the tag. Figure 1 shows a few possible shapes.

Figure 1 Sample shapes for tags

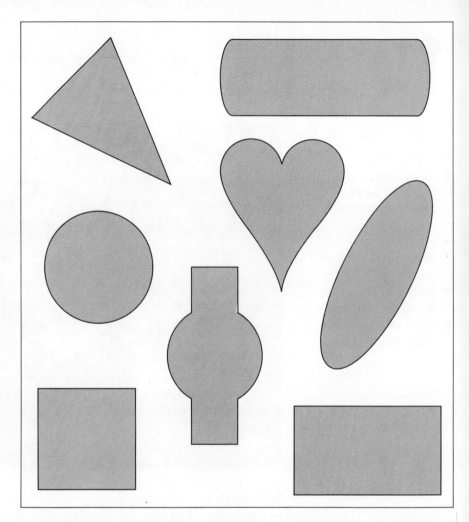

(c) What other information should the questionnaire consider? Possibilities include the material (e.g. plastic, wood or metal), the colour, the size, and the type of fastening. How might you be able to give interviewees an idea of these features in your diagrams?

(d) Having completed your range of designs, select the ones you are going to put forward with your questionnaire. You need not actually *make* the tags, but you should draw them carefully (preferably to scale).

Task 2

You should now be in a position to put your questionnaire together into a workable format. As you design the questionnaire, think about how you will make sense of and use the information − known in research as *data* − which you collect. This may help you decide what questions to ask, or how to word them.

Task 3

The next stage is operational planning:

(a) How many interviewers will there be?

(b) How many people do you need to interview to make your results worthwhile?

(c) To get a balanced view, will you need an equal number of females and males?

(d) How will interviewers approach interviewees – will you have a standard introduction to explain the research?

(e) When will be the best time to conduct the interviews?

(f) Where would be the best places?

(g) Will you need more than one day to carry out your research?

(h) Do you need to ask interviewees their names?

(i) Have you considered costs?

Task 4

Conduct the interviews.

Task 5

When you have completed your interviews, analyse your findings. How is this to be done?

Produce a chart that illustrates your data. Can you draw any conclusions from the research?

Task 6

Present a report to the manufacturers, advising them of your findings. Inform them of:

(a) the size of the *sample* (the total number of people you interviewed), and the numbers of males and females;

(b) when and where the research was carried out;

(c) which tag design proved to be the most popular;

(d) which material was most popular;

(e) whether there a colour preference;

(f) whether there was a preferred fastening;

(g) how much people might be prepared to pay.

————IMPORTANT————

Read these information pages:

A Social-survey techniques

B Better by design

12 Job shadowing

_____AIM_____

To develop your
- observation and recording skills
- ability to collect and organise information
- skills in working in a group

Note: This assignment continues the work done in Assignment 8.

Introduction

This assignment uses a special technique for finding out in detail what is involved in carrying out a job. *Job shadowing* involves being with a worker for a period of time, usually a day, observing all they do and recording what you have seen. Job shadowing provides a very valuable opportunity for you to see in detail the type of work you could be involved in and the skills which you would need in a chosen craft area.

If you have done Assignment 8 you will already have identified a firm and a contact within it, possibly a manager or someone in charge of personnel, and made arrangements for a job-shadowing exercise. If not, your first task is to arrange to shadow someone.

Task 1

In a group, agree the objectives of the job shadowing. Be sure that the observation and recording that everyone does will be similar, so that you can draw it all together after the visits. Make a list (as in

Figure 1 The agreed objectives in job shadowing

Observation	Recording

Figure 2 Details about the craftspeople shadowed

Key skills used	How often?

Range of different tasks seen

Equipment and tools used

Figure 1) of what you agree about observing and recording targets for the job shadowing.

Task 2
Carry out the job-shadowing exercise in the way agreed.

Task 3
Meet as a group and compare your experiences. Collect together details from your reports, listing each craftperson's skills, the range of work and the type of equipment and tools used (Figure 2).

_____IMPORTANT_____

Read this information page:
H Behaviour in
    manufacturing
    workshops

Task 4
In your group, discuss what you have seen and learnt. Make recommendations about activities or experiences which could be included in your programme to give you opportunities to practise some of the skills and tasks you have seen in job shadowing.

13

High-tech help

———AIM———

To develop your
- written and oral communication skills
- skills in presenting information graphically
- awareness of the role of computers in small businesses

Introduction

The word 'crafts' brings to mind products that are hand-made and of high quality, and this might seem at odds with the idea of computers. Computers, after all, represent the new automated, robot-produced high-tech world with standard products produced on an assembly line. However, many craftspeople are using new developments in IT to make their businesses more successful. Craft-based businesses have the same needs as any other businesses, and can also take advantage of the rapid developments that have been made.

This assignment gives you the opportunity to find out more about computers and the roles that they can play in craft-based activities. You will be visiting a number of craft businesses and assessing the extent to which they use, or are planning to use, computer technology.

Task 1

Using your local paper, *Yellow Pages* and so on, contact a selection of local craftspeople. Whether you write or telephone, explain that you are trying to identify three or four who are currently using computer technology, and three or four who are seriously considering doing so. Add also that you would like to visit them personally, and explain your assignment.

Task 2
Make a checklist for yourself of the information you will want to collect during your visits to these businesses. You will need to know the basic details about each firm, such as its name, address, telephone number and main activity. For those already using computers, ask about the sort of equipment owned, its uses, costs, advantages and disadvantages. For those seriously considering using computers, ask about the proposed uses and the amount that the firm is thinking of spending.

Read Tasks 4 and 5 to see how you will present this information.

Task 3
Visit the businesses who have responded favourably, and collect the information as planned.

Task 4
Collate your information as a chart, using the example in Figure 1 as a guide.

Figure 1 Chart showing uses of computer technology

Firm	Activity	Equipment	Cost	Advantages	Disadvantages

Task 5
Write a letter to each of the firms involved in your study, thanking them for their help. Include a summary of your findings, but leave out the names of the firms. You might comment on the range of prices, the disadvantages experienced or the types of uses.

Task 6
Prepare a wall chart which shows your results visually. You might include photographs of the more interesting uses that you found; pictures of computers and new equipment; lists of advantages and disadvantages; and diagrams showing the proportions of firms using the equipment for the different purposes. You *may* also be able to add letters from firms saying how useful your assignment has been to them, in helping them to think about their own equipment needs!

_____IMPORTANT_____

Read this information page:
A Social-survey
   techniques

14 Housework

―――――AIM――――――

To develop your ability
- to make considered choices
- to calculate costs
- to make comparisons of quality items

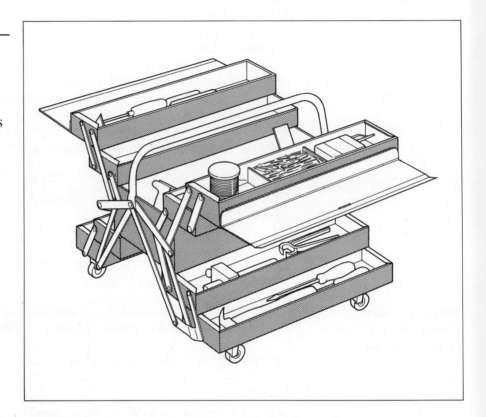

Introduction
Everyone from time to time has to undertake tasks which call for the use of tools, whether simply to repair items around the home or in connection with their job. Over a lifetime the amount of money spent on tools will be considerable. Some people have a systematic approach to tool purchase; others simply buy individual tools to do particular jobs.

In this assignment you are given a case study. You are asked to consider which tools are necessary for doing up a house that has been neglected.

Task 1
The tool list in Figure 1 was drawn up by a do-it-yourself (DIY) home owner after purchasing a new house (Figure 2).

Make a list of the items from Figure 1 which *you* consider should be purchased first. Explain how you made your choice.

Figure 1 Tool list

1 variable - speed power drill (10 mm)	1 set of screwdrivers (plain)
1 single - speed jig saw	1 set of screwdrivers (cross)
1 orbital - action sander	1 set of AF/metric combination spanners
1 circular saw (140 mm)	1 machinist's vice (medium)
1 saw table	1 AF/metric socket set (medium)
1 drill stand	1 set of nail punches
1 six - piece twist drill set	1 pair of mole grips (medium size)
1 set of chisels (for wood)	1 scriber (small)
1 set of chisels (for metal)	1 pair of combination pliers (small)
1 hand saw (560 mm)	1 pair of combination pliers (medium)
1 tenon saw (250 mm)	1 claw hammer (medium)
1 wood plane (jack) (50 mm)	1 trimming knife
1 multi - purpose ladder	1 steel rule
1 blow lamp	1 tool box
1 set of allen keys	

Figure 2 Description of the house

The house is 34 years old. It has been rather neglected by an elderly couple who in their later life could not afford the cost of repairs. The couple died; and the present owner has just purchased the house after a 16-month period during which the house was empty. The water was turned off during the winter but even so there was some damage to the water pipes. The electricity system was in need of complete overhaul; this has already been done by a professional electrician. The outside paintwork was very poor and in places the window frames have rotted; in one bedroom damp is present. The water that got in has rotted the floorboards near the centre of that room. The garden was fairly well looked after, but needs to be tidied.

Task 2

Cost each item in Figure 1, and find out the lowest and the highest price for each item. Find out and report on why the prices differ. You may find that a tool catalogue or a visit to a supermarket will help. Also, don't forget to ask people you know who have purchased tools recently.

Task 3

The home owner decides to purchase all of the items in Figure 1 over a five-year period. One or more tools will be bought every three months. List the items the owner should buy on each occasion.

———**IMPORTANT**———

Read this information page:
F Care of tools

Task 4

Look again at Figure 2. Has the owner chosen the *right* tools? Are there any that you feel are missing from the tool list? Explain why you feel the missing items should be included.

15 Second edition

To develop your skills in
- investigating and collecting information
- working effectively in groups
- oral communication
- practical and manipulative tasks

Introduction

Whatever the area of manufacture, whatever the production process, whatever the type of commodity being made, there is always the possibility of improvement. A product could be improved by making it:

(a) more efficient in carrying out its job;
(b) easier to produce;
(c) cheaper to produce;
(d) more attractive;
(e) closer to the wishes of the customer.

Of course, many of these areas are inter-related and their importance varies with the nature of the product. Hand-made products, for example, tend to be expensive because of the time involved in their production — a tailor-made suit will always be more expensive than a suit bought ready-made from the same material. In this case (e) is more important than (c). But it is still worth the company investigating ways of making the tailor-made suit more easily or more cheaply.

Large manufacturing companies often have whole departments given over exclusively to product research and development (R & D). A major British manufacturer (of components and electrical goods) recently reported annual spending of £40 million in R & D alone. This level of spending is necessary to keep ahead of the competition and to ensure that the products match customer demand and the availability of new materials. Even very simple products are constantly updated and remodelled — the tin-opener, for example, has developed considerably in the last thirty years.

The purpose of this assignment is to enable you to explore the ways in which a number of products have developed, why they have developed in this way, and what you could do to develop them still further.

Task 1

In groups of four, put together a list of products whose history you are going to plot. You list should include:
(a) a piece of electrical equipment (e.g. the television, the vacuum cleaner, the washing machine);
(b) a piece of mechanical equipment (e.g. the car, the motorbike, an industrial machine such as a lathe or press);
(c) a hand-crafted product (e.g. tailor-made clothes, hand-made furniture, specialist equipment);
(d) a personal item (e.g. the pen, the pocket calculator, the watch, the personal stereo).

Identify these four items as precisely as possible: for example identify a named brand of washing machine, or a particular type and make of car.

Task 2

Allocate one product to each member of the group for the research project. Your task is to identify:
(a) the changes that have taken place in the product in the past twenty years (you may have to reduce this period for some products — e.g. pocket calculators haven't been around for twenty years yet);
(b) the reason in each case for these changes;
(c) any areas in which the product could be improved.

There are many ways in which you can collect the data which you will need. You could begin near home with some of these products, by asking parents or friends whether they still have any older versions. For most people a washing machine or vacuum cleaner has to last quite a long time. You could also contact the manufacturer, by letter if the company is not local, to ask for help in identifying major product changes over the years.

Task 3

Having collected the information on changes in your particular product, the next task is to present your findings to the rest of group. You will do this as a 10-minute oral presentation (Task 4), but you should prepare a chart or diagram of your findings to help you and to act as a visual aid. (Figure 1 shows a sample chart.)

Figure 1 Chart of a product history

Product description	Product change	Year	Reason for change

Task 4

Make a 10-minute oral presentation to the others in the group on your main findings.

Task 5

Now consider how the products you have looked at could be improved. Study the product histories you have been given. How could the item be made more efficient, more effective, less expensive?

As a group, take 5 minutes for each item and brainstorm ways in which it could be improved. One member of the group should list the suggestions on a piece of paper.

Now allocate to each member of the group a specific product. Consider the suggestions made, agreeing the ones that are practical and identifying what problems the company might have in carrying out these improvements.

Present these as a short report identifying the realistic developments and any disadvantages or difficulties there may be.

———IMPORTANT———

Read this information page:
B Better by design

16 Gummed up

_____ AIM _____

To develop your skills in
- undertaking practical experiments
- recording information and presenting it graphically

Introduction

Joining and bonding similar or different materials are common activities in any craftsperson's work. Developments in adhesives technology over recent years have provided the craftsperson with an extensive range from which to select when creating or repairing items. Many products on the market have remarkable strength and some act in a way that makes the adhesive process irreversible.

There are dangers to be avoided when handling a number of adhesives. Some give off overpowering vapours; others may damage eyes and skin. The adhesive industry has been anxious to work with Government Health and Safety officials both in making users more aware of the problems and in co-operating with retailers in the prevention of solvent abuse.

This assignment will provide you with an opportunity to look at adhesives from a practical viewpoint, as well as increasing your knowledge of the wider issues.

Task 1

In the craft area that you have been studying, what are the processes that require the use of adhesives? What are the adhesives that are used? List your examples and rate how effective you think the adhesive is at doing its job — score '1' for 'excellent' down to '5' for 'poor'. Use the chart in Figure 1 as a model.

Figure 1 Use and effectiveness of adhesives

Craft process	*Adhesive used*	*Rating*				
		1	2	3	4	5
Fixing formica veneer to wood surface	Thixofix (thixotropic)					

Task 2

One question that should always be asked by the adhesive user is 'Will it hold?' In perfect conditions, with the right adhesive for the job, a permanent join can be achieved — often stronger than the materials being joined. You may have seen television advertisements showing people and cars suspended from supports that have been cut and then joined with powerful impact adhesive.

Just how strong *are* adhesives? Arrange to use the school or college's Newton gauge for measuring force (Figure 2). Make a number of simple wooden blocks for the experiment and obtain a range of adhesives. You will need to screw cup hooks into the blocks to attach them to the Newton gauge and the weights.

Use a variety of times for adhesive contact, longer and shorter than the times recommended by the manufacturers. Record the breaking strain at which the join fails (Figure 3).

**Figure 2 Using a
Newton gauge**

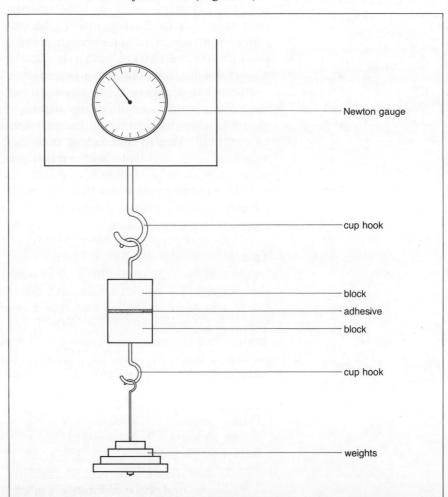

Newton gauge

cup hook

block
adhesive
block

cup hook

weights

**Figure 3 Record of
breaking strain**

Adhesive	Time	Breaking strain	Comments
	3 minutes	1 kg	slow-setting

Task 3

Many adhesives have properties that make them dangerous if not handled as stated in manufacturers' instructions. Potential hazards are numerous:

- burns
- skin irritations
- skin bonding
- blinding, or eye irritations
- intoxication by fumes
- risk of fire
- risk of explosion

Carry out an investigation of the potential health and safety hazards when working with popular adhesives. Design a poster for workshop use which identifies each adhesive, advises users of any potential dangers and recommends appropriate precautions.

———IMPORTANT———

Read this information page:
C Health and safety

17 Work-experience project: costings

———AIM———

To develop your skills in
- identifying and estimating costs
- planning and carrying out field research
- interviewing and communicating with others

Introduction

Knowing how much it costs to make something is crucial to the efficient running of any business. Much depends on knowing the detailed costs, which determine not just the profits made but also the very survival of the company.

It is intended that you carry out this assignment during the work experience you undertake as part of your course. In completing the assignment you will learn how a firm identifies and calculates the costs of making its products, and how it arrives at its prices. In doing so you will learn about the different ways of calculating costs. You may even help the firm *improve* its methods of costing goods!

Task 1

Before carrying out assignments during your work experience you need to prepare and plan the work you are going to do in some detail, and to get the approval of the work-placement manager. First read through this assignment briefing and complete all the preparatory work. Then write to your contact at the work placement, arranging a visit before the work experience begins. You will need to explain the nature of the assignment and what you are planning to do.

Task 2

Identify the kinds of hand-made products that are made at your work placement.

Choose one to investigate in detail. Make a three-dimensional drawing of this product.

Task 3

By observing how the product is made, find out the following information about it:

(a) What materials and components are used to make it?
(b) What tools and equipment are used to make it?
(c) What processes does it go through (planing, sawing, etc.)?
(d) Which employees are involved?
(e) How is it packaged?
(f) How is it transported to its point of sale?
(g) What other staff and processes are necessary but not directly involved in making it (accounts staff, storekeepers, etc.)?

When you have collected all this information, present it in the form of a flow chart illustrating how the product is made.

Task 4

Make a list of all the items of cost involved in making the product. Don't forget that some of the costs will be hidden, for example heating and lighting, and rent and rates for the workshop.

When you have made a list, present it in the form of a chart like the one in Figure 1.

Figure 1 Costs of production

Item of cost	Amount
1 hour of a skilled carpenter's time	
6 planks of light pine (150 mm x 25 mm x 1 m)	

Task 5

Having carried out your basic investigation, arrange an appointment with the owner or manager of the company to find out how the product is costed by the company, and how its selling price is arrived at.

Make a detailed list of questions you want to ask and carry out the interview. It may be useful to practise some of the interview questions with someone else before you interview the manager.

Task 6

After the interview, compare how the manager costed the product with your own findings from the earlier tasks. Are there costs which the firm should include but doesn't? Do you think that the company is charging the 'right' price for the product?

Make a written report about the costing of the product and send it to the manager. This report may also help you present your findings to your group when your work experience is completed.

————IMPORTANT————

Read these information pages:
H Behaviour in
   manufacturing
   workshops
I Costs

18 Convenience foods

──── AIM ────

To develop your ability
- to carry out instructions
- to assess value for money
- to compare methods of production

Introduction

'Convenience foods' are a familiar part of many people's diet. They are designed to reduce the amount of preparation time for a meal, or to make the task of producing it simpler. Many people welcome convenience foods, for these reasons and others; other people say that only traditional ways of preparing food give the right taste, texture, colour, and so on.

There are a lot of strong *opinions* for and against convenience foods. This assignment will give you an opportunity to look at the *facts*: to make comparisons by setting up tests, and to reach conclusions about the convenience foods and traditional preparation methods you have studied.

Task 1

There are hundreds of convenience foods on the market so you will need to be selective. The products the assignment will focus on are flour-based packet mixes, such as sponge mixes, scone mixes, and cheesecake mixes. (You may decide to look at a different product which fits better with your overall programme in craft-based activities or which interests you more.)

Find out what is available for sale locally and select two or three different mixes. Note the contents of each packet and any ingredients *not* included in the packet. Work out the cost of the ingredients. Lay out your findings in a chart like the one in Figure 1.

Figure 1 Table of costs

Packet cost	Packet mix	Ingredients	Additional ingredients	Full cost

Task 2

Find recipes for the same kind of foods. Old-fashioned recipe books might be worth looking at for that 'traditional flavour' that many people talk about.

Compare the recipe ingredients with your packet mix and make notes on the differences. Can you explain these differences? For example many packet mixes will use dehydrated ingredients or contain preservatives to increase the packet's shelf-life.

Task 3

Work out the costs of making the same quantity of cakes, scones and so on by traditional methods, starting with the raw materials, as your packet mixes will make. Compare the costs. What are the different requirements for tools and equipment for the two methods?

Task 4

Costs are one thing, but how do the two kinds of food compare for looks and taste?

Make up batches of products by each method and test them on volunteers. Decide beforehand on the questions you want to ask tasters, and make a checklist you can use in tests (e.g. choice of 'like'/'dislike' or a scale of 1−10). Will you want to create any special conditions for the tasting test? For example would it be wise to blindfold the volunteers?

Task 5

Analyse your results. Are there any overall preferences?

Ask yourself questions about *producing* the products. Which method was easier? Which took longer? How easy to follow were the instructions on the convenience pack compared to the recipe?

_____IMPORTANT_____

Read these information pages:

A Social-survey techniques

I Costs

Task 6

Write a report on what you have found out. Can people *really* tell the difference between 'home-made' and convenience foods in tests? Who could benefit from using the packet mixes you have looked at? How convenient *are* convenience foods?

19 On the safe side

AIM

To develop your
- skills in presenting information simply and clearly
- awareness of the need for safety
- decision-making skills
- ability to recommend safe practice

Figure 1 Injuries at work

Figure 2 Major parts of the body

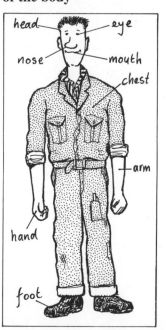

Introduction

All accidents are unexpected and unplanned, and for many people accidents have tragic long-term consequences. The effects of an accident are not necessarily confined to the person or people directly involved: usually others, such as family and friends, are affected.

Accidents can and must be prevented. If each of us takes time to make sure that we have acted in a safe manner, fewer accidents will happen. The overall aim must be an accident-free environment.

This assignment will help you to become more aware of the need for safety and to learn how to act for your own safety and that of your colleagues.

Average causes of industrial injury each year in England and Wales			
25.2%	Handling and lifting goods	7.2%	Being hit by moving transport
20.0%	Using moving machinery		
16.3%	Falling	7.1%	Using hand tools
9.2%	Striking against or stepping on objects	8.0%	Others, including accidents involving electricity
7.0%	Being hit by falling objects		

Task 1

First read Figure 1, which shows the kinds of accident which happen to workers.

Then look at Figure 2, which shows the main parts of the body. Make recommendations to reduce the likelihood of an injury in each area, similar to those in Figure 3.

Figure 3 Avoiding injury

HEAD

Keep hair short

Long hair which is not fitted underneath a cap or protective safety hat is a major cause of injury. It can become entangled in a moving machine, and if it hangs down in front of the eyes then there is a greater risk that the worker will strike against an object.

Wear head protection

Most manufacturing and construction firms now insist that visitors wear head protection as there are now more low-level structures, designed to use all the work space available.

Figure 4 Part of the Health and Safety at Work Act

Section 7. It shall be the duty of every employee while at work:
(a) to take reasonable care for the health and safety of himself and of other persons who may be affected by his act or omissions at work.
(b) to co-operate with his employer or [Health and Safety] representative so far as is necessary to enable a duty or requirement to be performed or complied with.

Task 2
The Health and Safety at Work Act places a responsibility on all people at work and states duties of employees. Figure 4 contains an extract from the Act.

Re-write Figure 4 in a form which you find easier to understand. Present a copy to all the people in your group at school or college.

―――IMPORTANT―――

Read this information page:
C Health and safety
H Workshop safety

Task 3
Any behaviour which distracts a person's attention, such as practical jokes and fooling around, can cause accidents.

Make a list of safety rules which could be given to new workers in a small craft workshop. Highlight the correct behaviour for people at work.

20 Setting up in business

_____ AIM_____

To develop your
- understanding of the process of setting up in business
- ability to seek out information
- knowledge of organisations that can assist commercial ventures

Introduction

In this assignment you are going to consider just a few aspects of setting up a craft-based activity: machinery, tools, raw materials, premises and financial assistance. The various tasks are designed to give you ideas and information which will be helpful. You could expand the tasks to study other considerations if you wish.

Task 1

What kind of craft-based business would you like to set up?

If the business is to be successful you need to consider three things:
(a) your interests;
(b) your abilities;
(c) your experience to date.
You might find it helpful to make lists of these things.

Task 2

Consider these questions.
(a) What equipment will you need to start off your business (machinery, hand tools, hand power tools, work benches, etc.)?
(b) Who are the suppliers of this machinery, these tools, and so on?
(c) What raw materials will you need?
(d) Where will you get these raw materials?
(e) What costs will be involved just to get going?

Task 3

(a) You are going to need premises for your operation. Are you going to rent or buy? In groups, consider these two options and decide which you would choose. Make notes of the advantages and disadvantages both of renting and of buying.
(b) How much space will you require?
(c) What services will you need (e.g. an electricity supply)?

Task 4

Now you have decided on the kind of premises you want, are such premises available in your area? How can you find out? Here are some suggestions:
(a) estate agents;
(b) Industrial Development Officer at your local council;
(c) the 'commercial properties' section in your local newspaper;
(d) enterprise agencies.

These people will need to know the size of premises you want and what kind of operation you would like to run. *You* will need to find out:

(a) how much the premises will cost if you buy;
(b) how much the rent will be if you lease;
(c) how much the rates will be;
(d) which services are already laid on;
(e) whether there are any other charges you need to know about;
(f) the location and details of access to it;
(g) whether there are any special arrangements for somebody setting up a craft-based business for the first time.

It would be preferable to visit people personally. If you cannot, telephone or write to get the information.

When you have acquired a list of premises, decide which one would best suit your intended business.

Task 5

You now have information which relates to the costs of your premises and the machinery and so on that you need. Where can you get *financial* advice and assistance?

There are various financial agencies which offer assistance; for the purpose of this assignment, though, concentrate on the main banks. (Your school or college probably has existing links with one or more of these banks.) Nearly all of the banks run small-business services and publish materials to help people who want to set up in business on their own. They will tell you about business development loans, and other loan schemes for small firms, leasing arrangements, how much money you yourself would need to provide, and so on.

In groups, discuss the information you have received from the banks. Decide which method of financial assistance you would choose for *your* business.

Task 6

Draw a flowchart that shows clearly the route you would follow now to set up your chosen business.

—————IMPORTANT—————

Read these information pages:

D Using flowcharts
J Sources of
   information

Information pages

A

Social-survey techniques

Uses

Social surveys in the form of questionnaires and public-opinion interviews have become a common technique of gathering information both in industry and by social scientists.

Purposes
In industry and commerce, survey techniques are used to monitor customer reaction to, and opinion on, a wide variety of issues – for example the colour and shape of packaging or the effectiveness of a particular advertising campaign.

Kinds of survey
The most common forms of survey techniques used are:
(a) *promotional campaigns* – for example the distribution of products to a selected area in order to assess customer opinion, or the use of free samples for the same purpose;
(b) *questionnaires* – these are used for gathering more detailed information about peoples' reactions or opinions; and
(c) *interviews* – this survey technique is not common in industry or commerce, and is mainly used when detailed or selective information is sought.
In this information page we will concentrate on the use of questionnaires.

Usual format

The public-opinion questionnaire can be used for a variety of purposes such as the assessment of peoples' political views or social attitudes, as well as to discover product preferences.
   The questionnaire consists of a series of structured questions designed to draw from the interviewee specific answers.
   To achieve this structure in a questionnaire, it is necessary to follow a certain procedure. The procedure can be divided into six stages.

Stage 1: Survey design
(a) You must first decide what it is you wish to find out.
(b) Then you have to decide on the number and type (for example, the age, sex and occupation) of people who you

think will be able to supply the required information, or be representative of such groups.

(c)   Next you must choose the wording and order of the questions. The main problems to watch out for at this stage are these.

- *Wording* It is possible with questionnaires to word a question in such a way that you get the answer you expected. For example the quesiton, 'Do you think the Government should go to war to defend the liberty and freedom of its subjects?' would probably draw from interviewees a positive response to going to war. Whereas a question such as 'Should the Government go to war on behalf of another country?' would probably draw the opposite reply.

- *Bias* This is reflected not only in the wording used, as in the example above, but also in the way questions are constructed. By quoting figures of authority in questions, the interviewer is probably introducing bias into that question. People tend to respect authority and would not normally wish to be seen in contradiction to it. For example, 'Would you agree with the opinion of the majority of doctors that smoking should be banned in public places?' (rather than 'Do you think smoking should be banned in public places?').

- *Ambiguity* It is important to avoid the use of questions that are vaguely worded and that can therefore be interpreted in more than one way. For example, 'Are you in favour of government policies that affect old people?' is ambiguous because a 'yes' answer could be given for two reasons. The interviewee could mean that he or she is in favour of active policies in general, even though he or she feels that the existing policies affect old people negatively. The answer would also be 'yes' if the interviewee felt such policies affected old people positively.

- *Selecting questions* Here it is important to decide on the nature of the response you are seeking. In questionnaires there are two main types of questions: *closed questions* restrict the answer that can be given to either a 'yes' or 'no' answer, or to placing a tick in the appropriate box; *open questions* allow interviewees to answer in their own words. Each type has advantages and disadvantages. On the whole, closed questions are much easier to analyse but much more difficult to construct. In most questionnaires a mixture of both types is used.

- *Personal questions* In a general questionnaire, try to avoid too many personal questions, as people do not like answering such questions. If you do need to ask them, try to keep them indirect (closed questions are best for this). For example:

'How old are you?'
please tick box:

☐ under 18

☐ 18 to 35

☐ 36 to 55

☐ over 56

- *Order of questions* When designing a questionnaire, try to group questions of a similar nature together. This not only makes the progression more logical when answering, but helps later at the analysis stage.

Stage 2: Sample survey

In order to assess the soundness of a questionnaire a sample survey is normally conducted. This entails trying out the questionnaire on a proportion of the total sample.

As your time will probably be limited, we suggest that you try your questionnaire out on a few fellow students to see if it contains any structural problems. If it does then you will need to make the necessary alterations before conducting your full-scale survey.

Stage 3: Conducting the survey

There are three basic ways in which a survey based on a questionnaire can be conducted:
(a) selecting a sample group of people and then sending them a questionnaire by post, asking for completion and return;
(b) stopping a selection of appropriate people in the street and asking them to fill in your questionnaire there and then;
(c) the same as (b), but with the interviewer asking the questions and writing the answers on the questionnaire form.

The type of procedure you use will depend on the sorts of questions you are asking; people do not liked to be asked too many personal questions in the street but they might answer if the questions are sent to them privately through the post.

Each method has its own advantages and disadvantages. The main problem with method (a) is that there is normally a very high non-response rate. This could mean that your survey becomes biased because only a selective group of people respond.

Stage 4: Classifying the information

Once the survey has been conducted, some analysis of the results is necessary. Before this analysis can take place, the answers to the questions have to be classified; in other words, you have to examine the replies you have received and categorise them under headings or titles. With closed questions this is easy because all you have to do is count up the number of ticks or 'yes/no' answers. Categorising can sometimes be made easy by using the actual question in the survey as the title for your classification.

Stage 5: Data analysis

Once you have classified your data it is then possible to analyse it – that is, you will be able to look at the results and see clearly the conclusions you can draw. What these conclusions will be depends largely on the type of questions asked and the people interviewed.

Stage 6: Presenting your results

It is important to present your conclusions clearly and logically. This can be done by the use of statistical data, by graphical presentation, or by presenting them verbally or in a written report. The choice of presentation will depend on:
(a) the type and nature of questionnaire; and
(b) the nature of the audience being informed of your results.

For information on the presentation of statistical data, see information page D.

Here is an example of a questionnaire on present and future UK energy resources.

This survey's aim is to discover the public's awareness of present and future UK Energy resources: to find out the public's attitudes and reactions to the energy crisis.

OCCUPATION: _____

SEX: ☐ Male   ☐ Female

AGE: ☐ 16–20   ☐ 31–40   ☐ 51–60
         ☐ 21–30   ☐ 41–50   ☐ 60+

---

1. Which form of energy do you think generates the greatest proportion of electricity into the National Grid?

   ☐ Gas   ☐ Nuclear
   ☐ Oil   ☐ HEP (Hydroelectric power)
   ☐ Coal   ☐ Others

2. Which of the following do you think consumes most energy in this country?

   ☐ Iron and steel industry   ☐ Domestic
   ☐ Industry (others)   ☐ Others
   ☐ Transportation

3. Are you in favour of nuclear energy production?

   ☐ YES   ☐ NO   ☐ UNDECIDED

4. Which of the following 'alternative' energy sources have you heard of?

   ☐ Wind   ☐ Geothermal
   ☐ Solar   ☐ Plants
   ☐ Ocean Thermal   (eg Water-Hyacinth)
      Energy Conversion   ☐ Splitting hydrogen
   ☐ Tidal   ☐ Others _____

5. Which of the previous list do you consider would be the most realistic future alternative energy source for the UK?

   _____
   _____
   _____

6. Do you consider that the following are vital to UK energy production?

   Natural Gas   Coal   Oil
   ☐ YES   ☐ YES   ☐ YES
   ☐ NO   ☐ NO   ☐ NO
   ☐ UNDECIDED   ☐ UNDECIDED   ☐ UNDECIDED

7. Do you have central heating?   ☐ YES   ☐ NO

   If YES please tick one of the following
   ☐ Oil   ☐ Electricity
   ☐ Gas   ☐ Other _____
   ☐ Solid fuel   _____

8. Have you ever seriously considered the use of Solar Panels at home?

   ☐ YES   ☐ NO
   If YES, what did you eventually decide and why?

   _____
   _____
   _____

9. Do you consider that an energy conservation policy should be an
   important consideration of our Government?
   ☐ YES   ☐ NO   ☐ UNDECIDED

10. Have you taken any steps towards conservation in your own home?
    ☐ YES   ☐ NO
    If YES please tick which of the following

    ☐ Loft insulation   ☐ Draught-stripping
    ☐ Cavity wall insulation   ☐ System controls
    ☐ Hot water tank insulation ☐ Double-glazing

    ☐ Others _____
    _____

11. Did you receive a local government grant for any of the above?
    ☐ YES   ☐ NO

B

Better by design

The need for design

We are surrounded by the products of our industrial society. Few of these products are made the way they are just out of common sense and public demand. Almost every one has been carefully designed.

Before beginning the expensive work involved in production of a new item, the manufacturer will want to see accurate and detailed plans, a prototype of the item, and a report on the usefulness, safety, and potential sales of the item. The manufacturer will also want to know about relevant legislation and probable consumer response to the item; and he or she will want a full costing.

The nature of design work

Designing is always a mixture of things. The designer will usually be given an outline of what is wanted, but this may be only in a brief letter. There may be complex laws that affect the production, packaging and sales of the commodity. There may be special requirements for overseas use – electrical appliances may need different voltages, for example, and instructions may need to be translated into other languages. Servicing will be very important, as will safety.

As well as satisfying all of the technical requirements, the designer must make sure that the product is attractive. This will involve getting a number of things right: the feel, the weight, the colour, the overall design and the efficiency and the reliability. Reliability may be tested by making a prototype – a single experimental version of the new item, made using the planned materials. This can be tested by experts and potential users, and any necessary modifications made before mass production begins.

Bad design or no design will lead to low-quality products which do not perform well.

The job of a designer

There are many different sorts of designers, each working in a particular way and in a particular area of industry. Visual designers are concerned with communicating through signs, symbols, forms, colours and the relationship between these. Industrial designers are concerned with the design of functional objects, and with the economics, techniques and materials involved. Graphic designers work in the world of publishing, in advertising, in books and in the press. (A graphic designer will have been involved in the production of this book.) Research designers experiment with design itself and develop new areas of design.

The best designers have some of each of these talents, but most companies employ teams of designers, each working in a specialist area.

What does a design look like?

Designs come in many different shapes and sizes. The design for a new town hall might take the form of a detailed report describing the materials and structure, accompanied by drawings and a small-scale model. The design for a book cover would be drawn and painted to look like the final cover; the designer might show variations on clear overlay sheets.

Designs can be in many formats: what matters is the quality. The design should be clear and accurate and should leave nothing out.

Good design

Good design is partly a matter of opinion. Everyone has his or her own preferences in terms of colour, shape and feel. People choose different clothes, for example, and it would be a dull world if they did not.

Consider a chair. If all that were needed were something to sit on, an old box or crate would do. But people want their furniture to look good, to feel good and to be comfortable and safe. A well-designed chair would meet all of these criteria. In practice, there are already hundreds of different chair designs, and there will probably be hundreds more.

People will never agree that one chair design is better than all the others, but they will probably all agree that some are good and some are bad. And the ones that are good will be good not by chance but by design.

C

Health and safety

A case study

Janet worked in a department store. One day she was asked to go to the stockroom to fetch replacement stock for her department. The box she wanted was shelved a little higher than she could reach. At the far end of the stockroom was a movable step-ladder, but Janet thought that if she stood for a moment on the bottom shelf she could just reach the box she wanted. In fact the shelf gave way, however, and Janet fell backwards. She needed stitches for a cut on the back of her head, and her right ankle was in plaster for almost two months.

Janet acted foolishly: she should not have used the shelf as a ladder, especially as the steps were available. But did she know that? Did the department store have a code of conduct for the stockroom? Had Janet been trained in the correct way of collecting stock from the shelves? Was it her job anyway?

No employee should behave in such a way as to put himself or herself, or anybody else, at risk of injury. On the other hand employers should have codes of safe practice, they should provide training for their employees and they should provide whatever equipment is necessary to protect their employees from injury.

The Health and Safety at Work Act 1974

Health and Safety at Work
etc. Act 1974

CHAPTER 37

LONDON
HER MAJESTY'S STATIONERY OFFICE
Reprinted 1975
£1·50 net

The Health and Safety at Work Act lays down the responsibilities of employers and employees in the workplace. Legislation before 1974 had tended to relate to particular industries (e.g. mines, quarries, and chemical plants): it did not cover *all* workers. The 1974 Act updated existing legislation, widened its scope to include all workers, and set up the Health and Safety Executive to make sure that the regulations in the Act are carried out.

If the safety and health measures described in the Act are not followed by employers, the employers can be heavily fined and may even have their businesses closed down until they do carry out the measures. On the other hand, if an employer provides safety equipment and trains workers in safety procedures but the employee fails to use the equipment or ignores the safety regulations, the employer cannot be held responsible for any injury incurred by that worker. Indeed failure by an employee to carry out his or her employer's safety policy can in some instances lead to dismissal.

Sources of information

The Health and Safety at Work Act 1974 is a long and detailed document. You will be able to find a copy in a public library. Parts of the Act probably relate to your school or college, and the school or college should have a safety policy – ask if you can see it. Copies of all government legislation are available from Her Majesty's Stationery Office, but these may be expensive. You can get particular sections of the Health and Safety at Work Act from the HMSO.

Under the terms of the Health and Safety at Work Act, trade unions have the right to elect safety representatives in the workplace. You can get information from individual trade unions or from local Trades Councils on their policies on Health and Safety.

The Workers Educational Association (WEA) runs courses on health and safety for Union safety representatives – the WEA could be another source of information. If you undertake any work experience the provider will instruct you early on in the firm's health and safety policy, telling you about protective clothing and equipment, the safe use of machines, fire drills, and so on. When you visit any works ask about health and safety procedures and equipment.

Making it work

The Health and Safety at Work Act on its own cannot prevent accidents or provide safe workplaces. What it does is to provide legal protection and a statutory framework. But in everyday work it is employers and employees who make the Act effective by being conscious of health and safety in the workplace and by co-operating to prevent accidents and keep to safe working practices.

Using flowcharts

What is a flowchart?

A flowchart is a sequence of instructions shown as a diagram. It is one way of breaking down a complex task into its component parts, including any choices that must be made. Figure 1 shows an example.

Figure 1

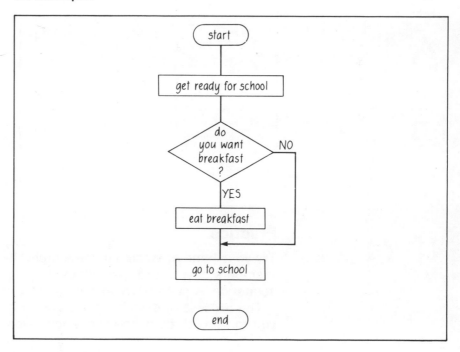

Conventions

So that other people can understand what is meant in a flowchart, certain conventions are followed in the use of shapes and the way they are joined together. It will help if you follow these as well.

Direction of flow

The main direction of flow is from the *top* of the page to the *bottom*, and from the *left* to the *right*. If a line flows up the page or from right to left, add an arrowhead to show this. Only one flowline should enter a box – if two lead to it, they should join before the box. See Figure 2.

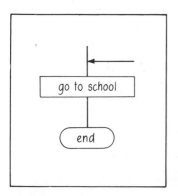

Figure 2

Shapes

Rectangular boxes are used for *statements* (often actions), and diamond-shaped ones for *questions* with a 'yes' or 'no' answer. (Each diamond should have *two* flowlines leaving it.) Parallelograms are used when *information* is taken in or given out. See Figure 3.

Beginnings and *ends* of the flow are shown as boxes with rounded ends (Figure 4).

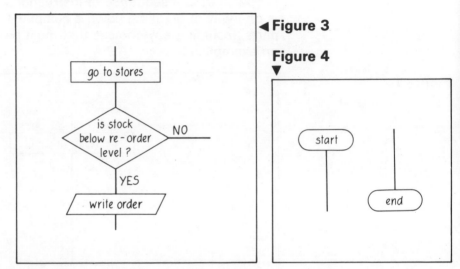

◄ **Figure 3**

Figure 4
▼

Practice

Try your hand at writing some simple flowcharts. Ask other members of your group to check that the boxes are in the right sequence. A good flowchart is easy for other people to follow.

The process of devising a well-organised flowchart will help you to think through the sequence involved in an activity.

Technical terms

In many jobs you will come across words and phrases that are new to you. You may also meet terms you thought you understood, now being given quite different meanings.

Vocational studies

In each vocational area there is a specialist vocabulary. You *need* to learn at least some of this. Tutors could remove technical terms from the teaching you receive, and work-experience providers could set you tasks in a way that avoided using them, but this would not really be kind to you as your teaching or work-experience would then be unrealistic.

You need to be attentive and curious about what's going on. Being able to use the correct names for tools, equipment or processes makes you feel happier in a new subject, and shows other people that you're committed to it.

Learning technical terms

Ask questions
Nobody will think the worse of you if you ask questions about technical terms. On the contrary, they will realise that you are taking an interest. When you hear technical terms used, try to see them written down – then any notes you make will be accurate and helpful. It is easy to mishear a term.

Use the terms
Help yourself to remember technical terms by using them in the practical situations in which they appear. If you are unsure, use a dictionary or reference book to double-check.

Keep your own list of the technical terms that are important in your vocational area. Add illustrations if these are helpful. You might like to write some of the terms on cards that you can carry in a pocket – you could look at these in spare moments. If other people are learning the same terms you could design a home-made card or board game in which technical terms have to be matched to pictures or definitions.

Jargon
Sometimes people use technical terms in unhelpful ways. They

may be trying to impress you or to bluff you into thinking they have a superior knowledge. Don't be baffled or feel put down – ask them to explain what you don't understand. You'll soon discover whether or not they know what they're talking about.

F

Care of tools

All craftspeople use tools of various kinds. Whatever the occupation, tools are used. Even a person sitting at a desk all day may use a pen, a stapler and a ruler.

You will find that over a number of years you will be spending a great deal of money on the tools of your trade. The care of tools will therefore become an important part of your skills as a craftsperson.

Whatever type of tool you are using you need to ask yourself questions all the time.

(a) Is the tool clean?
(b) Is the tool being used correctly?
(c) Is the tool damaged?

Below we look at each of these questions in turn and give the reasons you should ask them.

Is the tool clean?

It is extremely annoying to go to use a tool and find it dirty. This can put you into a bad mood before you start the job! Also it can suggest to other people who see or use your tools that you are a slovenly worker.

Dirty tools are dangerous, too, and may result in injuries to your hands and other parts of your body. There have been many reports of workers using dirty tools who received permanent eye injuries. Try to imagine what it would be like to have to live the rest of your life with only one eye – just because you had used a tool with grease on the handle, your hand had slipped, and you had damaged the retina in your eye. *Care now can prevent injuries later.*

Precautions

(a) Always clean tools before you put them away.
(b) Keep a clean rag at hand to wipe tools you have used.
(c) Do not fool around when using tools. Concentrate at all times.
(d) Keep your tools in a tool box.

Note that some tools – such as reamers and shears – need a light covering of oil to protect the cutting edge, but this oil should be in the correct place, the cutting edge, *not* on the handle.

Is the tool being used correctly?

The poor worker will always blame the tools when things go wrong. The good worker will take pride in the tools and only use them for the correct task. Some *do's* and *don'ts* will help you:

Do's
- Be prepared for the tool you are using to slip.
- Make sure that there is no damage to the tool.
- Keep tools clean.
- Clean tools before you use them.
- Keep cutting tools sharp and lightly oiled.
- Keep any workpiece firmly gripped in the vice.
- Choose the correct weight of hammer for the job.
- Use a ring spanner or a socket rather than an open-ended one.
- Change saw blades regularly.
- Ensure that the spanner is the correct size.
- Use the correct size of screwdriver.
- Use the *palm* of your hand (not your fingers) to turn a spanner or any other lever.

Don'ts
- Never use excessive pressure on any tool.
- Don't use anything on a tool to increase its leverage.
- Never use an adjustable wrench if a fixed type can be used.
- Don't force a spanner onto a nut.
- Don't use pliers to remove nuts.
- Don't use dirty tools.
- Don't use other tools as a hammer.
- Never use a screwdriver as a driftpin.
- Never use a file without a handle.
- Never use a screwdriver as a lever.
- Never hammer a screwdriver into a joint face.

Is the tool damaged?

The competent craftsperson will always keep an eye for damaged tools. Whenever you discover any damage you should either replace or repair the tool.

Most cutting tools can be rectified by sharpening – twist drills, reamers, chisels and punches are all tools which can be re-sharpened to bring them back to a useable condition. However special angles must be maintained: before you attempt this task, ask a skilled person to show you how. You should be aware that there are theoretical considerations concerning cutting angles, speeds, grinding processes and sharpening techniques. Your tutor will help and advise you during your training programme.

Tools damaged in other ways (that is, apart from their cutting edges) should always be replaced. Ill-fitting spanners, for example, have been the cause of many injuries. Broken files have resulted in cut hands and arms. Cracked hammers have shattered and sprayed splinters of metal around the workshop.

Watch carefully for any damage. If you *do* find a damaged tool, put it in the scrap bin – the only safe place for it!

Precautions
- Check that the hammer head is tight on the shaft.
- Ensure that the handles on files are tight and undamaged.
- Keep cutting tools in a special box or in their guards.

Choosing tools

Only a few tools are mentioned above. You may find that there are *hundreds* in your chosen trade! Try to relate the ideas given here to your own particular area of work. The basic principle is to give a great deal of careful thought to your tools. In return they will give *you* a great deal, by helping you to do your job better, more easily and more efficiently.

Keep in mind that there are some manufacturers who produce poor-quality tools which are very good imitations of high-quality tools. Many cheap brands would let you down; they could cause injuries to yourself or others. Good tools are expensive to buy, but over the years they give greater service. It is better to buy *one* really good tool each month than to buy cheap sets of tools of poor quality.

Take pride in *good* tools and they will give you good service.

Working in a team

Many of the assignments in this book require you to work with others in order to get the best results. Similarly, many jobs or recreational activities outside of school or college require teamwork. To get the best out of a team, everyone involved needs to know what the team is trying to achieve, and what part he or she has to play.

Pros and cons

Potential problems

(a) One of the biggest problems faced by a team is that it is made up of individuals – each team member will see things in a slightly different way.
(b) Some members will be more highly motivated than others, and this can create friction.
(c) The division of tasks within a team often results in one person getting what appears to be a more interesting or more important job than somebody else.

Advantages

(a) A team *can* act in a very effective way. It can produce results that are far better than those that could be achieved by several individuals all working on their own.
(b) Some tasks are very large, and one person just wouldn't be able to tackle them alone.
(c) A team can use the varying strengths of several individuals, and help each other with any weaknesses.

Examples of situations needing good teamwork

- A residential experience
- A Youth Theatre Group production
- An enterprise or mini-company project
- A Students' Union programme
- A course 'open evening'

Good teamwork

Communication
It is vital that there is good communication between team members. People need to have the same aims and be willing to

listen to each other. When a team is being set up, careful thought needs to be given to:

(a)   when the team will meet;

(b)   how information gathered, or tasks completed by individual members, will be shared with the team;

(c)   recording what has been done and what remains to be done.

Leadership

The direction in which the team goes, and the allocation of members to particular tasks, can be achieved in a number of ways.

In many situations in life and work the leader of a team is chosen by the team itself. Team leaders have a difficult role to play. They may be in the spotlight more than other members; they may be 'blamed' if things go wrong; they may face jealousy from some other team members. A team leader should try to resist any temptations to exploit his or her power as leader. In a well-organised team the team leader would look to the other members to provide a great deal of support.

Not all team arrangements need a leader. In small teams in which the tasks are fairly clear the division of work might be quite a simple matter.

In training situations in which a leader *is* needed it is a good idea to rotate the role. This way a number of team members can gain experience of leadership. The less glamorous jobs too can be rotated, so that nobody feels put upon or left out of the action.

Planning and organisation

If the work of a team is to be efficient then there needs to be careful planning and agreement about activity. It is very tempting for everyone to dash off and start doing their own thing, but this can produce chaos.

The team can begin planning by asking these simple questions: Who? What? Where? When? Why? How?

It is a good idea for the team to stop occasionally, to check how things are going. These questions may help:

(a)   What have we done so far?

(b)   Is it going the way we planned?

(c)   What changes are required to achieve the team's objectives?

You will find working in a team exciting, challenging and most worthwhile.

Many situations in adult life require experience and confidence in teamwork. In a special survey carried out to discover what employers valued in young workers, the ability to work as part of a team was listed among the top ten qualities employers were looking for.

Workshop safety

As you gain experience as a worker you will gradually learn how to behave safely in a workshop. Safe behaviour patterns will become instinctive, and will help you remain accident-free throughout your working life.

People are sometimes said to be 'accident prone'. There is no such thing! People have either a good behaviour pattern or a bad one. Safety is an attitude of mind and very personal. No one else will do your thinking for you: you yourself must learn to be aware of the dangers around you. When you begin work in a workshop, attention to safety will need a great deal of thought. Later safety will become part of your routine and you will automatically act safely. The following points are just *some* of the areas you should pay particular attention to.

Noise

There is always noise in workshops. To some extent you must learn to live with it. But you should try to be aware of any unusual noise or change in noise. A sudden change may indicate a danger: for example a fork-lift truck suddenly coming towards you or a chain-lift pulley about to fall from above. Never ignore another worker's call to beware of danger.

Keep noise to a minimum

Because noise is inevitable in a workshop you should try to reduce the noise *you* make. Whenever you talk to others in the workshop, use tools and equipment, operate machines, or move around the workshop, keep in mind the level of noise you are making. If everybody does this the overall noise level will decrease. This is safer at the time; it also helps to prevent deafness in later life.

Liquid

Liquids should always be kept in containers. There should *never* be liquid on the floor, as it makes the floor slippery. Someone could slip or fall, perhaps with tragic results.

If you spill anything, clean it up *at once.* Liquid left on the floor will penetrate into the floor and it will become impossible to remove it completely.

Movement

Do not hurry around the workshop: to do so is very dangerous. As a new trainee you may feel you should impress people by carrying out the task required quickly. Don't! In any manufacturing workshop there are dangerous machines, tools and equipment, such as sharp-edged cutting tools, saws, rotating machines and electrical tools. If you hurry you may fall or collide with such equipment, and the consequences could be tragic.

Go about your task in the workshop positively but carefully.

Skin care

Chemical substances are rapidly absorbed into the body through open wounds, but rarely through unbroken skin. You should therefore protect any wound with suitable dressings as soon as you notice it. Remove any liquid from the skin as soon as possible, even if you don't think the liquid is dangerous. Get into the habit of cleaning skin regularly after contact with liquids, otherwise you may forget and transfer the liquid to your mouth where it may enter your body.

Liquid on the skin can cause irritations or 'dermatitis' (a skin complaint which is difficult to cure). It is a good idea always to use a barrier cream as a protection against skin-damaging agents. The regular use of such a cream will also make your hands simpler to clean when your working day is over.

Smoking

For your own health, the best thing is not to smoke at all. If you *do* smoke, take care that the habit does not become a careless or dangerous one.

Even otherwise harmless material can become potentially dangerous near a naked flame. Make sure *you* are not the cause of a fire or an explosion in your workplace. You might not live to regret it!

Ensure that you smoke only in areas where it is allowed. Some workshops have a blanket *No Smoking* rule. Watch for notices, and don't ignore these rules, which are there for everyone's safety.

High spirits

Skylarking, joking and practical joking have *no* place in a workshop.

Costs

For all firms engaged in craft-based activities, it is essential to keep a close watch on how much an article or component costs to make. How much we sell something for is largely based on how much it costs to make.

Yet it is not always easy to take all the costs of production into account. The problem arises because there are many ways we can look at costs and measure them.

Ways of looking at costs

'How much did that item cost to make?' To answer this question fully we must include *all* the costs actually incurred in making the item.

Basically there are two types of cost: *fixed* and *variable*.

Fixed costs
Definition Costs which do not increase or decrease very much whether you make a lot of the item or only a little. (You can think of fixed costs as those costs you have to pay even if you do not make the item at all.)
Examples Rent and rates for the workshop; interest on money borrowed to buy equipment.

Variable costs
Definition Costs which *do* increase or decrease when you make more or less of the item. (You can reduce variable costs by reducing the amount of the item you make.)
Examples Costs of raw materials; wages; heating bills.

Total costs
Definition All the costs incurred in making something: the sum of the fixed costs and the variable costs.

Average costs
Definition The total costs divided by the number of things we make. It is sometimes called the 'unit cost' or the 'cost per unit'.
Use This is a very useful way of looking at cost. You can increase the unit cost by a given percentage (a 'mark-up') to pay for the skills needed to run the business. The marked-up cost is the price at which you sell each item.

Example of costs

Here are some of the important costs:
- rent
- rates
- heating
- lighting
- fuel
- raw materials
- interest on loans
- mortgage payments
- transport
- postage and telephone
- wages

Hidden costs

Some costs are hidden. Take the case of a man setting up his own business with £1000 of his own money. He doesn't actually pay anyone anything, but if he had left the £1000 in the bank or building society it would have earned interest. The lost interest is a hidden cost. *All* costs should be accounted for.

Keeping records

A golden rule for all businesses, large and small, is to keep *all* documents relating to buying and selling and to keep an accurate record of *all* transactions. Today small businesses can often afford small computers on which all this information can be stored as a database or combined as a spreadsheet.

J Sources of information

Information is a general term for the facts without which our complicated modern world would grind to a halt. There is now so much information in the world that it is impossible for any one person ever to know it all, or even to know where all of it can be found.

Those who have information (or access to it) may have power. But information itself has no value unless it can be used. Knowing how to find out what you want to know is therefore a very useful skill.

Sources of information

Information is all around us. For example, there are information centres in every town. There are libraries, tourist information offices, council offices, Citizens' Advice Bureaux, chambers of trade, police stations, bookstalls and newsagents, bus and railway stations.

Where do you start?
The largest source of information is the people in the community. People are usually very willing to help. For example, you may feel that your doctor is only available when you are ill, and your bank manager only when you want a loan, but they will usually be very helpful if you approach them with questions about their own specialist areas. The business community too is a rich source of knowledge, skills and experience.

If you don't know the names or addresses of local sources, the best place to start is a library. Your own school or college probably has a well-equipped library of its own; if not, try the local town library. The town libraries will probably have an information desk or an information section.

Getting information

Asking
One good way of finding things out is to ask other people. This may not be easy if you are shy, but if you can develop your confidence in asking for help you will find that many other things you do will become easier.

Looking

How much do you know about the firms that are trading in your town? Almost every side street will have some employer carrying out some commercial activity. As you walk around, start paying attention to these areas.

You may have an industrial estate near you – a place where several firms that need manufacturing or workshop space operate next to one another. Take a walk around one – you'll be surprised at the variety of companies there.

Reading

There are now many directories that list the names of firms and what they do. Your local library information section will contain a large number of such trade directories. The most commonly available is the *Yellow Pages* (Figure 1). Chambers of commerce and other business organisations produce lists of their members, as do many county councils.

Figure 1

Chambers of trade, chambers of commerce and chambers of commerce and industry differ in their national organisation, but they are all voluntary associations of businesses. Each will have a local secretary who will be a very useful contact if you want to get in touch with local firms. Local newspapers usually have special sections giving the names of local service and repair companies, and many local councils produce lists of local services.

Writing

You may want to visit, or get information from, a local organisation or firm. Business people are busy, and to begin with it may be best to write. Say clearly in your letter what you want to do or find out, and make the letter as neat as you can manage – the employer will get his or her first impression of you from your letter.

Most organisations are used to receiving inquiries from students, and they may send you more material than you expected.

Listening

If you arrange an interview, plan carefully so that you make the best use of the time. Write your questions down before the interview so that you don't overlook any area of interest to you. Pay attention, be serious and be polite. Take a notepad and try to record what you are told.

Good listening is a skill in itself. The person you are interviewing will probably give you lots of information: if you are nervous or waiting to ask your next question, you will miss much of what is said. To practise listening, try interviewing another member of your group about a hobby or a holiday.

Using computers

Computers make information storage, retrieval and communication much simpler. More and more local libraries, schools and even major shops are installing computer terminals which make information available at the touch of a few buttons. The information may actually be stored on much larger computers many miles away, but it will be sent to your terminal down the telephone line, or broadcast like television programmes, perhaps even via satellite. Talking, reading, writing and listening will continue to be important, but as the amount of information available continues to grow, the use of computers will become more and more common.